DATE DUE

Rain

Joy Palmer

C.1 1998
14.98

RSVP
RAINTREE
STECK-VAUGHN
PUBLISHERS
The Steck-Vaughn Company

Austin, Texas

Editor: A. Patricia Sechi
Design: Shaun Barlow
Project Manager: Joyce Spicer
Electronic Production:
   Scott Melcer
Artwork: Brian McIntyre
Cover Art: Hugh Dixon

Library of Congress Cataloging-in-Publication Data
Palmer, Joy.
      Rain / Joy Palmer.
         p.   cm. — (First starts)
      Includes index.
      Summary: Describes the phenomenon of rain, how and why it falls, how it can be measured and predicted, and its effect on the world and its life.
      ISBN 0–8114–3413–3
      1. Rain and rainfall — Juvenile literature.   [1. Rain and rainfall.]
I. Title.   II. Series.
QC924.4.P35 1993
551.57'7 — dc20   92–38554
                              CIP AC
Printed and bound in the United States

1 2 3 4 5 6 7 8 9 0 LB 98 97 96 95 94 93

Contents

What Is Rain?

Millions of tiny water droplets make up a rain cloud. These are all different sizes. Big droplets bump into smaller ones, and they join together and make even bigger droplets. When they are big and heavy enough, they fall to the earth. If the temperature is above freezing, they fall as raindrops. The larger the drops, the faster they fall. Beneath the clouds, we can feel the raindrops falling on us.

▽ Thousands of tiny droplets join together to make one raindrop heavy enough to fall.

Water in the Air

Sometimes rain falls and forms puddles on the ground. Heat from the sun dries up the puddles, leaving the ground dry. This is called **evaporation**. The water has turned into an invisible gas in the air, called water vapor. When **water vapor** cools down, it turns back into water droplets. This change from gas to liquid is called **condensation**.

▷ Fog is like a cloud on the ground. Water vapor in the air condenses into droplets of water.

▽ Puddles of water soon evaporate in the heat of the sun.

 Water vapor rises from a hot drink. As the water vapor cools, we see the droplets of water.

5

Around and Around

Air rises when it is warmed, or in order to move over hills and mountains. Water vapor is carried by the air from the land and sea, up into the sky. Higher, the air is much colder. This cold air makes the water vapor condense, forming tiny droplets. These droplets form the clouds in the sky. Raindrops fall out of the clouds and back down to Earth, and so the **water cycle** goes around and around.

▷ The water cycle never stops. Rain falls to Earth, turns into water vapor, rises, turns into droplets, and falls to Earth again.

▽ We can sometimes see rain pouring out of the clouds far away in the distance.

The water cycle

3 Clouds of droplets form and rain falls.

2 High up, the air cools and the water vapor condenses.

4 Rain wets the land and fills rivers and streams.

1 Rising air carries water vapor upward.

Measuring and Forecasting

Rainfall is easy to measure using a rain gauge. A gauge is made from a hollow cylinder containing a long, narrow tube. A funnel at the top collects falling rain and passes it into the tube. The amount of rain that falls is measured against a scale on the side of the gauge.

Cloud shapes are good signs which help to tell us if rain is on the way. Some old country sayings also tell us about signs of rain.

▷ Some people have their own ways of forecasting the weather.

▽ Each cloud shape has a name. Some of these clouds are a sign that rain is on the way.

cirrocumulus

cirrus

cumulonimbus

cumulus

A closed pine cone means rain.

An open pine cone means dry weather.

▽ Meteorologists study the weather. They use a rain gauge to measure rainfall.

Thunderstorms

Tall, dark cumulonimbus clouds usually bring a heavy rain storm with thunder and lightning. Lightning is an enormous spark caused by **static electricity** building up in a cloud. Eventually it is freed in a sudden flash of light. Flashes of lightning heat the surrounding air. As the heated air collides with cool air, crashes of thunder can be heard.

▷ Lightning can flash from one cloud to another.

▽ Lightning can flash between a cloud and the ground. It may strike buildings or trees.

Rainbows

We usually think of sunlight as being colorless. Scientists call it "white light." It is really made up of the colors red, orange, yellow, green, blue, indigo, and violet. Rainbows help us to see the colors of light. When sunlight shines through a raindrop, the water splits the light and colors are bounced off the droplet. The colors of light make a beautiful arc shape in the sky.

A thick cloud blocks the sun.

The rain is lighter now.

△ When sunlight shines on the spray from a waterfall, it forms a rainbow.

The clouds break up.

Sunlight shines through the rain. The raindrops split up the colors in light and make a rainbow.

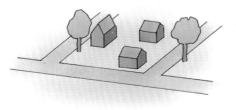

All About Raindrops

Most people draw raindrops to look like teardrops. If you could look at them before they fall on the ground, you would see that they are actually shaped a bit like hamburgers! As they leave their cloud, raindrops are round. They are fattened up by the air as they fall. Raindrops vary in size from 1mm to 6mm. When the air is full of drops smaller than 0.5mm, we call it drizzle.

▽ Big, black rainclouds are full of water droplets which join together and then fall to the ground as raindrops.

◁ Raindrops are shaped like flat-bottomed circles. They hang on leaves before falling to the ground.

◁ Waterproof clothes and umbrellas keep us dry in the rain.

Plants Need Rain

Most plants use the rain after it has soaked down into the soil. Plant roots and stems contain bundles of very small tubes. Water is pulled into and along these tubes. It then travels through the roots and stems, and into the leaves. Water evaporates from the leaves of plants into the air. The larger the leaves, the more water is lost by evaporation.

▽ Plants lose water through their leaves. A rubber plant has large leaves. It grows in wet rain forests.

◁ Cacti grow in hot, dry areas. They have tiny spines instead of leaves. These do not lose water as quickly.

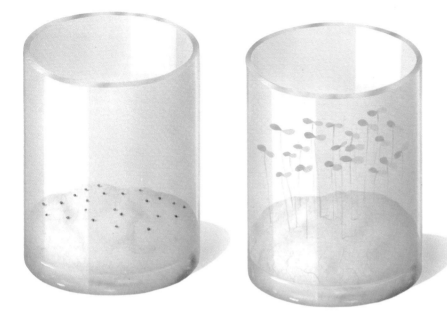

◁ Plants need water to live. Add water to seeds and they will begin to grow.

▽ The roots of some trees and plants are very long. They grow deep down into the soil to gather water.

Animals and Rain

Just like plants, all animals need water to keep them alive. Many wild creatures find a puddle, pond, or stream from which they can drink. Some creatures, such as fish and water insects, live in water all the time. Earthworms always come to the surface on a wet day because they would drown underground! If no rain falls, streams and ponds may dry up, leaving many animals to die.

▷ The animal called a hippopotamus lives in Africa. It spends most of its time in water.

▽ Ponds and streams are homes for fish, water insects, and frogs, and for visitors like toads and newts. They all depend on rain.

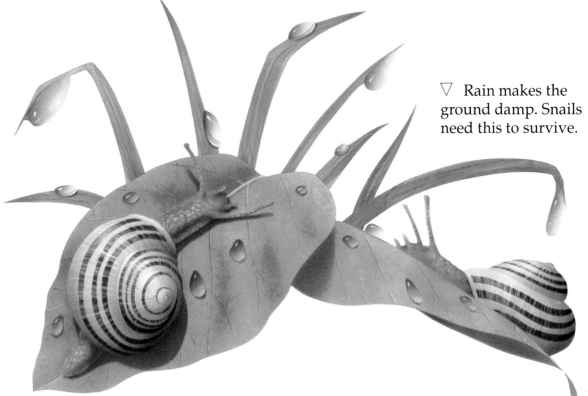

▽ Rain makes the ground damp. Snails need this to survive.

Around the World

Rain does not fall evenly over the Earth. Deserts are the world's driest places. In some deserts, it does not rain for years. The wettest place on Earth is called Mount Waialeale in Hawaii. It rains there nearly every day. Tropical rain forests have regular rain storms. Rain forests have a lot of sunshine as well, so the air is always warm and steamy.

▷ Many different kinds of plants and animals live in rain forests. There is warmth, water, and food all year round.

▽ There is such little rain in a desert that not many living things will survive there.

▽ Mount Waialeale is the wettest place on earth. It lies on an island in the Pacific Ocean.

Living with Rain

All living things need water. Without rain we would have no food or drink. Sometimes, people have to live with too much rain. A very heavy storm or days of heavy rain may cause a flood. Rivers and lakes can then overflow and may flood the countryside and streets, covering cars and houses. In countries where heavy rain falls regularly, houses are often built on stilts.

▷ Floods cover areas of land, which would normally be dry, with water. Too much water can damage buildings and crops.

▽ Indonesia often has very heavy rain. Many people build their homes on stilts to avoid flooding.

Rain on the Ground

When rain falls, it collects in oceans, lakes, rivers, streams, and puddles. In towns, it disappears into drains at the side of the road and is carried away along underground pipes. In the countryside, rainwater runs along hills and slopes to form streams and rivers. Some rain soaks into the ground on the way. The more it rains, the faster the streams and rivers flow.

▷ Rain falls on mountains and hills. Some water trickles underground, and some flows downhill in streams. These join together to become rivers which flow into the sea.

▽ Rainwater runs down a sloping roof into pipes. These carry the water down to the drains.

roof

gutter   drainpipe

Using Rain

We depend on rainwater for growing crops for our food. We also use it to drink. Water that collects in specially built **reservoirs** is eventually supplied to our homes. Sometimes rainwater collects underground. Wells are built from which water can be taken.

Rain also gives us opportunities to take part in water sports on streams, rivers, and lakes.

▷ Many people enjoy sailing on reservoirs and lakes.

▽ A dam holds back water to form a reservoir. Reservoirs collect and store rainwater. In dry years, the water level may be very low.

△ We use water to
wash and cook our
food.

▷ We all need
water to stay alive.

Acid Rain

Poisonous gases and particles are being sent out into the air all the time. These come from car exhausts, power stations, and factories. When these mix with water droplets in the air, the chemicals in them change. The raindrops turn into an acid liquid called **acid rain**. Acid rain can kill trees and plants, damage buildings, harm wildlife, and poison lakes and streams.

▽ Acid rain is caused by poisonous gases from cars, factories, and power stations.

▷ Acid rain is so strong that it can damage statues and buildings.

▽ Trees suffer very badly from acid rain, particularly on hills and mountains where rainfall is heavy.

Index

Photographic credits: Bruce Coleman Ltd. (J Shaw) 5, (H Reinhard) 17, (P Ward) 19, (N Bevore) 21; Crown copyright reproduced with the permission of the Controller of HMSO 9; Chris Fairclough Colour Library 27; Frank Lane Picture Library (H Binz) cover, 11, (K Ghani) 12, (Silvestris) 29; Robert Harding 3, Hutchison Library (T Beddow) 6, ZEFA cover, 14, 23.